***How the book is laid out*** – The book
Region, information about 29 selecte
species that regularly occur here duri
likely to be seen. Each site description
it, some of the birds it supports, and
routes. Other features are also noted

## BIRDING SITES

The birding sites cover all the main habitat types found in Odessa Region: the sea, coast, wetlands, steppe and woodlands. Unfortunately, Odessa lacks mountains or rocky areas. Most of the sites are in natural areas with beautiful landscapes and high biodiversity interest, not just for birds but also plants, insects (especially beetles, butterflies and dragonflies), reptiles and amphibians, and mammals such as spotted susl, *Spermophilus suslicus* and mound-building mice *Mus spicilegus* which are prey for several species of raptors. If nothing else, they are great places for a walk and picnic!

## MAPS

The maps are intended to give an impression of the site geography and principal habitats. They are not intended for accurate navigation. Various routes are indicated but in most places there are many alternatives to explore. Outside of protected areas and settlements, access to the countryside is generally open and almost every field or lake margin will have a dirt track that can be walked or driven along (sometimes a more comfortable ride than the road itself). The only warning is to be careful after rain as the silty surfaces tend to become soft and slippery; during dry weather they become very dusty!

**Key**

| | |
|---|---|
| | Black Sea |
| | Freshwater, fishpond |
| | River, channel, canal |
| | Reedbed |
| | Arable, pasture, open land |
| | Wooded area |
| | Settlements |
| ▬ ▪ ▬ | International border |
| ▬▬▬ | Surfaced road |
| •••••• | Suggested routes (foot, vehicle or boat) |
| ▬▬▬ | Railway |

## PLACE NAMES

The spelling of place names used in this Guide follows that used in Google Maps, but without apostrophes. A large part of Odessa has been incorporated in various empires through history and settlement names were given in Greek, Latin, Turkish, Russian and Ukrainian among other languages. During the Soviet period (1949 to 1991), all names were transliterated into the Russian Cyrillic alphabet, and often changed for political reasons. Today, there is a process of restoring older names, or giving new non-Soviet names, and converting all names to the Ukrainian Cyrillic spelling. These names are then transliterated to Latin script according to Ukrainian conventions. For example, a "new village" called Satu Noua in Romanian became Novoselskoye in Russian and is now Novosilske in Ukrainian. All these names can appear on maps and in online searches.

## CHECKLIST

The bird checklist comprises the 223 species which the author has personally encountered at least once during the last ten years of intermittent birding in Odessa Region, and considers the most likely to be seen by visiting birdwatchers. The overall Ukrainian bird list contains about 420 species.

# About Odessa Oblast (Region)

With an area of around 33,310 km$^2$, Odessa Region is the largest of the 24 Regions in Ukraine – it occupies about 5.5% of the country's territory, and is a little larger than Belgium. It extends along the lower Danube to the Danube delta, around the Black Sea coast over 300 km to Tiligul Liman, and curves inland some 250 km. The region also has 1,362 km of international borders with Romania and Moldova.

The population is just 2.4 million people, nearly 40% of whom live in the city of Odessa alone, so the countryside is quite thinly populated. The main economic activities are agricultural production and processing, petrochemicals, tourism and import/export transportation.

The northern part of Odessa Region is located in the hilly forest-steppe zone while the southern part lies in the flatter coastal Pontic dry to arid steppe zone. Black soils (chernozems) predominate, with good to high agricultural potential. Agricultural land occupies about 2.7 million ha (80% of the region) of which some 2.1 million ha (62%) are used for arable crops, especially for commodities such as winter wheat, corn, barley, millet, rapeseed and sunflower. These are grown in large fields (up to 600 ha), bounded by planted wind breaks. In addition, a wide variety of market crops, fruit and walnuts are grown. Agriculture in many parts of Odessa is limited by water availability, and closer to the Black Sea there are major irrigation networks drawing water from rivers, lakes and subterranean sources. Almost all of the water for Odessa city comes through a pipeline from the Dniester River.

The lower reaches of the main rivers (Danube, Dniester) and Black Sea coastal zone feature a range of high natural value wetlands (fresh and saline), steppe and riverine forest ecosystems, which have national and international significance. Accordingly, there are 124 protected areas and objects, with a total area of 159,976 ha (of which 150,843 ha, or 4.5% of Odessa Region, comprise nature reserves). Most of these sites are fairly small, dedicated for rare species or habitats, which are of particular interest for scientists. However, four of the protected areas are very large, namely Tiligul Liman Regional Landscape Park (sites 4 and 5); Lower Dniester National Nature Park (sites 11 and 12); Tuzly National Nature Park (sites 14 and 15); and Danube Biosphere Reserve (sites 18 and 20).

## ODESSA CITY

Odessa was founded in 1794 by Catherine the Great. By 1820, Odessa had become the most important commercial, industrial and cultural centre in the southern part of the Russian Empire and the greatest seaport on the Black Sea. Today, Odessa is the third largest city in Ukraine and the largest situated on the Black Sea. The city's wealth, derived from its trade and agricultural production, enabled it to create a "Pearl of the Black Sea". This was combined with a founding population from all over Europe (especially France, Italy and Spain) who fostered the independent, cosmopolitan values and religious tolerance of the city's inhabitants. It is remarkable for its broad tree-lined avenues, Renaissance-style architecture, places of worship, numerous city parks and of course the sea front. It is particularly proud of its famous landmarks such as the Opera Theatre, Potemkin Steps and Vorontsov Palace.

Where and when to look

Although this book recommends 29 sites for birdwatching, from the heart of Odessa city to the remote reaches of the Danube Delta or the steppe zone around Tiligul Liman, these are far from the only places where birds can be seen. Compared with much of Europe, it will soon be apparent to the travelling visitor that Ukraine is a vast country with significant areas of pristine natural habitats where birds (and other wildlife) can and do thrive.

## Around Odessa
Birding can start in Odessa city itself: thanks to its tree-lined avenues, quiet courtyards with vines and walnut trees, parks and the beach. Here, Syrian woodpeckers, black redstarts, tree sparrows and even long-eared owls find nesting places, while swifts and occasional hobbies zoom overhead. Either side of Odessa are large limans – coastal lagoons formed where rivers enter the Black Sea. These shallow, fresh to brackish waters attract large numbers of migratory waders, while fringing reedbeds host herons and paddyfield warblers.

## Northern steppe-meadows and forests
Farther north are great rolling valleys, holding meandering rivers and grassy slopes with scattered bushes of Russian olive and scattered copses of poplar and willow – perfect for harriers, rollers, bee-eaters, shrikes and buntings. These valleys lead to the southern edge of the Boreal forest where black kites, booted eagles, steppe buzzards and goshawks hold sway. Eastwards, the steppe zone begins in earnest, with little owls, calandra larks, Isabelline wheatears, and corn buntings everywhere.

## Rivers and coastal wetlands
South-west of Odessa, across the Dniester River and down to the Danube, is a region often called by its historical name of Bessarabia. Along the Danube, between Izmail and Reni, are five enormous floodplain lakes that support such large numbers of waterbirds that they are listed as wetlands of international importance under the Ramsar Convention. The narrow stretch of the Danube between Orlivka and the Macin Mountains in Romania is a favoured crossing point for soaring birds on migration: pelicans, storks and raptors in steadily moving spirals pass overhead in late April and early September.

## Wind-breaks are good for birds, too
When the steppe zone was brought into cultivation from the 1960s, lines of drought-resistant trees were planted to reduce soil blow and moisture loss. These belts form linear forests along the roads, used by birds such as golden oriole, Syrian woodpecker and long-eared owl but above all by rooks that have established huge rookeries extending many kilometres. These in turn support substantial colonies of red-footed falcons.

## The birding year
As elsewhere in Europe, the best period for birdwatching is from mid-April to mid-June when the migrants are moving through, summer breeders are setting up their territories, and the trees and reedbeds are reasonably transparent. From late July, the southward migration of waders begins, followed by passerines and raptors during August until mid-October. The best winter period is from mid-January to the end of February when tens of thousands of greater white-fronted geese, mingled with red-breasted geese, feed on the emerging winter wheat. Flocks of finches draw in sparrowhawks and merlins, and fish ponds are targeted by white-tailed eagles. Indeed, the winter can be a very rewarding time for birding in Odessa!

## General map of sites

The locations of the 29 sites covered in this Guide are indicated in this map. It can be seen that they are largely distributed around the perimeter of the province. This reflects the presence there of the marine and coastal zones, as well as the wetlands and mouths of rivers, where large concentrations of birds occur.

# List of sites

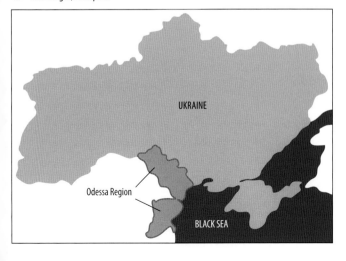

8

# ❶ Savranskyi Forest

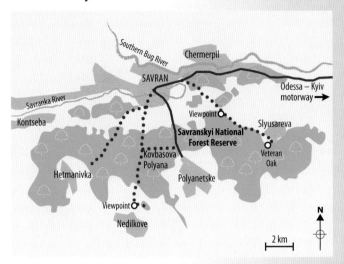

**STATUS** National Forest Reserve

**AREA** 8,397 ha

**FEATURES** The largest single block of native Boreal forest in Odessa, Savranskyi is dominated by pedunculate oak and ash, with some elm, lime and cherry. The forest is managed for timber production so it contains little dead wood and few trees over a century old. A grand veteran oak (said to date from 1425) is the main feature of a short ecotrail from the forest office in Slyusareva. The area holds a wide range of woodland birds, especially raptors and woodpeckers.

**GOOD FOR** Kestrel (top left), Nightjar (bottom left); Honey Buzzard, Black Kite, Steppe Buzzard, Booted Eagle, Middle-spotted Woodpecker, Hawfinch, Yellowhammer

**BEST PERIOD** May–September

**ACCESS** Best approached from Savran, where several dirt tracks lead to and across the forest. There are numerous forest tracks for walking (vehicles not permitted). The veteran oak is at 48.0939°N, 30.2034°E and viewing points at 48.1147°N, 30.1551°E and 48.0520°N, 30.0646°E are good for scanning the forest canopy and adjacent hunting areas for raptors.

## ❷ Tiligul River, Berezivka

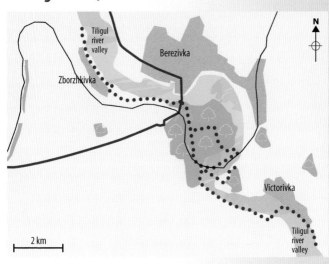

**STATUS** None

**AREA** ca. 1,130 ha

**FEATURES** The site comprises a block of managed mixed deciduous woodland (oak, ash, lime, cherry) lying in the floodplain of the lower part of the Tiligul River, with adjacent reedbeds and meadows; the river is generally a series of pools, flowing only after heavy rain. The juxtaposition of these habitats leads to the presence of a good diversity of birds in a relatively small area. The elevated banks of the floodplain allow scanning across the site.

**GOOD FOR** Wryneck (top left), Yellow Wagtail (bottom left); Steppe Buzzard, Great Reed Warbler, Icterine Warbler, Penduline Tit, Hawfinch

**BEST PERIOD** April–June

**ACCESS** The P55 road from Odessa to Berezivka goes through the centre of the site. Tracks off the road near the forestry centre (47.1914°N, 30.9096°E) lead upstream to Zborzhkivka, or downstream towards Victorivka. There is a bridge allowing access south of the railway. The tracks in the forest and lower part are not accessible for vehicles.

# ❸ Petrivka Forest

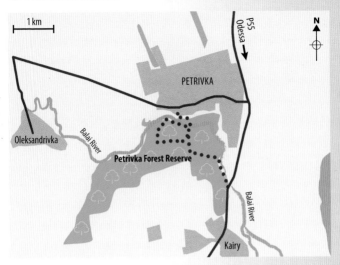

**STATUS** National Forest Reserve

**AREA** c. 420 ha

**FEATURES** The site comprises a block of managed mixed deciduous woodland (oak, ash, lime, cherry) as well as patches of conifers. It rises from the floodplain of the Balai River up to and along a ridge. As a reserve forest it is managed for timber production so it contains little dead wood and few old trees. There are wet meadows and small pools alongside the river channel, which is rather small and has irregular flow.

**GOOD FOR** Robin (top left), Wood Warbler (bottom left); Steppe Buzzard, Great Reed Warbler, Icterine Warbler, Raven, Hawfinch

**BEST PERIOD** April–June

**ACCESS** Petrivka village lies by the P55 from Odessa to Berezivka. Just before the road crosses the Baila, there is a lay-by on the forest edge with a picnic area (46.9503°N, 30.9666°E). From here, tracks lead into the forest, or along its margin. Another access point is from Petrivka (46.9645°N, 30.9492°E), where a footpath goes over the river into the forest. No vehicles allowed in the forest.

## ❹ Tiligul Liman, Kairy Bay

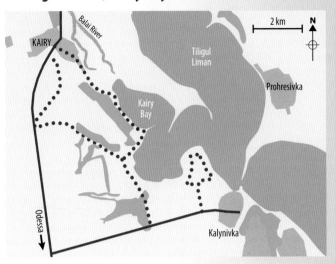

**STATUS** Lies within Tiligul Regional Landscape Park; Ramsar Site; Important Bird Area 092

**AREA** ca. 1,350 ha

**FEATURES** The site is where the Baila River enters Tiligul Liman, though its flow is so low that the area is mainly sand flats with patches of halophytic vegetation. The bordering slopes are good for steppe plants, and shrubby plantations provide shelter for passerines.

**GOOD FOR** Calandra Lark (top left), Montagu's Harrier (bottom left); Waders, Wood Lark, Red-throated Pipit, Tree Pipit, Isabelline Wheatear

**BEST PERIOD** April–June; August–October

**ACCESS** Kairy village lies on the P55 from Odessa to Berezivka. A track leads out east from the village (46.9381°N, 30.9637°E) along the southern edge of the bay. It reaches sandy cliffs with colonies of bee-eaters and rollers. Tracks at 46.9160°N, 30.9546°E lead up to fields and steppes with Calandra Larks. The promontory can be reached from Kalynivka.

## ❺ Tiligul Liman, Koshary

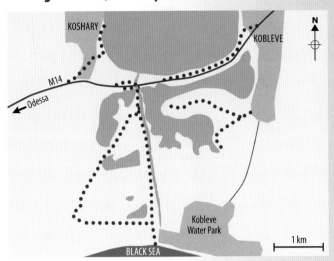

**STATUS** Tiligul Regional Landscape Park; Ramsar Site; Important Bird Area 092

**AREA** ca. 1,100 ha

**FEATURES** A large, elongated liman, bordering on Mykolaev province. It comprises stretches of shallow open water, sand bars, halophytic steppe and scrubby patches as well as the beach and views of the Black Sea. It is mainly a place for wetland birds, although the steppe area is good for larks, pipits and wagtails and the bushes support shrikes, chats and warblers

**GOOD FOR** Paddyfield Warbler (top left), Red-necked Phalarope (bottom left); Great Egret, Montagu's Harrier, Kentish Plover, Broad-billed Sandpiper, Terek Sandpiper, Marsh Sandpiper, Slender-billed Gull, Red-throated Pipit

**BEST PERIOD** April–June; August–October

**ACCESS** The M14 from Odessa to Mykolaev runs on a causeway across the site. There is a dense network of tracks all around the area (e.g. at 46.6546°N, 31.1672°E; 46.6668°N, 31.204°E; 46.6547°N, 31.1639°E; 46.6555°N, 31.1520°E). These are usually accessible by car, depending on water level. The embanked track leading south through the wetlands to the coast is good for Paddyfield Warbler.

# ❻ Kuyalnik Liman, Novokubanka

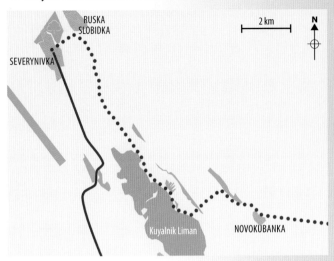

**STATUS** Important Bird Area 090

**AREA** ca. 2,040 ha

**FEATURES** The top end of Kuyalnik Liman comprises a thin sheet of very salty water, sand flats, and halophytic vegetation on its floor. The slopes are clothed with natural steppe, and the odd patch of false acacia plantation. Some of the steeper slopes are bare, and there is a sand quarry, that provides nesting sites for Sand Martins, Rollers and Bee-eaters.

**GOOD FOR** Bee-eater (top left), Corn Bunting (bottom left); Montagu's Harrier, Kentish Plover, Roller, Crested Lark, Isabelline Wheatear, Raven

**BEST PERIOD** April–June; August–October

**ACCESS** Perhaps the most remote place close to Odessa, reached only on dirt tracks, but well worth the effort! From the junction at Novokubanka (46.7297°N, 30.6653°E) head north to 46.7471°N, 30.6485°E and turn west along the field boundary, through the woodland and down the steep steppe slope to 46.7471°N, 30.6296°E. Proceed north along this track, turning west at Ruska Slobidka (46.8319°N, 30.6014°E) to reach Severynivka.

# ❼ Lower Kuyalnik Liman, Korsuntsi

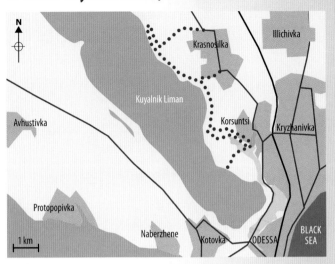

**STATUS** Important Bird Area 090

**AREA** ca. 3,550 ha

**FEATURES** Kuyalnik is the most saline of the Black Sea limans in Odessa Region, with beautiful salt encrustations decorating the mud around its edge (there is a famous spa on its western shore offering mud treatments). There is a small reedbed at the lower end. The slopes and cliffs have a sparse cover of steppe, and clumps of Russian olive and false acacia. At the top of the site is a small spring where birds gather to drink and sandpipers feed. There is a small reed-fringed freshwater lake at Korsuntsi which usually adds several more birds to the day list.

**GOOD FOR** Kentish Plover (left); Shelduck, Pied Avocet, Oystercatcher, Bearded Tit, Savi's Warbler and various gulls.

**BEST PERIOD** April–June; August–October

**ACCESS** There is a fairly dense network of dirt tracks along the eastern shore, mainly accessed from Korsuntsi or Krasnokova (e.g. 46.5879°N, 30.7629°E; 46.6281°N, 30.7605°E; 46.6816°N, 30.7434°E). Take care of driving close to the shore after rain.

# ❽ Odessa City Park

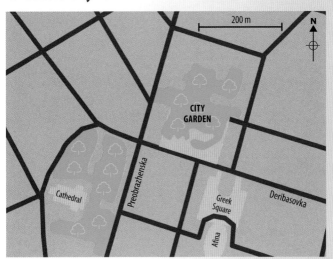

**STATUS** None

**AREA** 0.25 ha

**FEATURES** Relaxing over a morning coffee or evening beer is no excuse to stop birding! The mature trees planted in the city park are part of a woodland band stretching from the opera house, through the university grounds to the Greek Orthodox Cathedral. Many migrants stop off here, Caspian Gulls loaf on the buildings, and Hobbies often come over from Shevchenko Park to chase House Martins overhead in the evening.

**GOOD FOR** Black Redstart (top left), Syrian Woodpecker (bottom left); Hobby, Wryneck, Pied and Spotted Flycatchers

**BEST PERIOD** April–May; September–October

**ACCESS** From Deribasovka, Preobrazhenska or Lanjeronovska streets.

# ⑨ Odessa Shevchenko Park and Coast

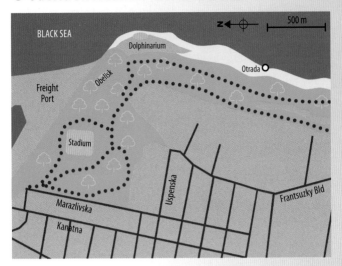

**STATUS** None

**AREA** ca. 1,200 ha

**FEATURES** The plateau and slopes behind the Odessa seafront south of the port are planted with a range of hardy ornamental trees and shrubs. The main part of the park was once a barracks and quarantine area for the port; today it mainly features the Chernomorits football stadium and various Soviet-era monuments. Outside of the main tourist season, there are several fairly quiet places where birding can be quite fruitful, especially for autumn migrants. The bay area is a good place to find seabirds. There is a small resident population of introduced Red Squirrels *Sciurus vulgaris*.

**GOOD FOR** Long-eared owl (top left), Red-backed Shrike (bottom left); Black-throated Diver, Black-necked Grebe, Levantine Shearwater, Eider, Hobby, Arctic Skua, Thrush Nightingale, Whinchat, Pied, Collared and Red-breasted Flycatchers, Goldcrest

**BEST PERIOD** April–June; September–October

**ACCESS** From Marazlivska and Uspenska streets. Some internal park roads are opened for limited vehicle access (paid entry in the summer).

# ⑩ Odessa Botanic Garden

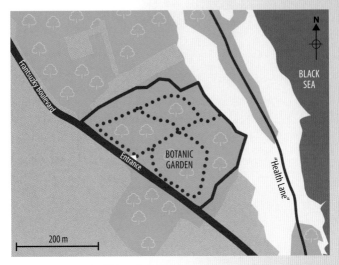

STATUS Private grounds of Odessa University

AREA ca. 70 ha

FEATURES This small but rather charming botanic garden, with a range of mature trees and shrubs, is situated on the cliff overlooking the southern coast of Odessa. It forms part of a more extensive wooded area along Frantsuzki Boulevard, but is a little haven of peace. It attracts a wide variety of migratory passerines, especially chats, flycatchers and warblers. There is a viewing point overlooking the sea which is handy for seabird searching. This is probably the best place in the city to practice some bird photography.

GOOD FOR Collared Flycatcher (top left), Redstart (bottom left); Syrian Woodpecker, Hobby, Thrush Nightingale, Pied and Red-breasted Flycatchers

BEST PERIOD April–June; September–October

ACCESS From Frantsuzki Boulevard (46.4402°N, 30.7684°E). There is a small entry charge (about €1.50) and permission is needed to wander around unguided. Please keep to the paths.

# ⑪ Lower Dniester National Park (by road)

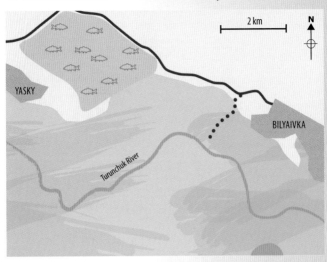

**STATUS** National Nature Park, Ramsar Site; Important Bird Area 091

**AREA** ca. 1,400 ha

**FEATURES** The site is mostly a vast reedbed through which the Dniester River, and its last tributary, the Turunchuk River, meander casually into the Dniester Liman, and adjacent fish ponds. The river levees support a narrow belt of willows and poplars, and there are patches of open water. Parts of the reedbed are harvested for thatch. There is a strong population of Wild Boar *Sus scrofa* that are occasionally seen around dusk.

**GOOD FOR** Marsh Harrier (top left), Greylag and White-fronted Geese (bottom left); Pygmy Cormorant, Little Bittern, Great Bittern, Black Stork, Glossy Ibis, Red-breasted Goose, White-tailed Eagle, Bluethroat, Savi's Warbler, Paddyfield Warbler, Bearded Tit, Penduline Tit, Great Grey Shrike

**BEST PERIOD** April–June; September–October

**ACCESS** This site can be viewed from the road between Bilyaivka and Yasky, which also passes a large fish farm. At 46.4979° N, 30.1700° E there is a track to a willow-lined embankment that goes deep into the reedbeds and provides excellent viewing opportunities.

# ⑫ Lower Dniester National Park (by boat)

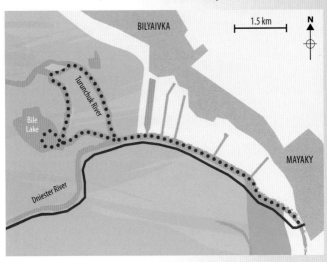

**STATUS** National Nature Park, Ramsar Site; Important Bird Area 091

**ROUTE DISTANCE** ca. 16 km

**FEATURES** The "site" follows the Dniester River and its last tributary, the Turunchuk River, as they meander through a vast reedbed before entering the Dniester Liman. Some of the river levees support a narrow belt of riverine forest comprising willows, alders and poplars and the route includes a visit to the waterlily-covered Lake Bile (the white lake).

**GOOD FOR** Grey Heron (top left), Night Heron (bottom left); Red-necked and Black-necked Grebes, Great White Pelican, Pygmy Cormorant, Little Bittern, Great Bittern, Ferruginous Duck, Hobby, Wryneck, Thrush Nightingale, Bearded Tit, Penduline Tit

**BEST PERIOD** April–June; September–October

**ACCESS** Boat trips can be arranged out of Mayaki from April to October; further details available from Salix Ecotours (see p. 1).

# ⑬ Akkerman Fort, Belgorod

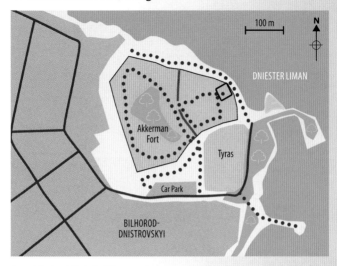

**STATUS** National Historical Monument

**AREA** ca. 200 ha

**FEATURES** Akkerman Fort and its neighbouring Greek settlement of Tyras are among the most impressive archaeological sites in Odessa. Situated at a strategic point on the western shore of the Dniester Liman, the fort serves as a rocky resting point for migrants coming in from the Black Sea. Its walls also provide holes and clefts for breeding Little Owl, Hoopoe, Swift and Pied Wheatear, as well as colonies of House and Tree Sparrows. The adjacent shore and liman proper are used by pelicans, waders, gulls and terns.

**GOOD FOR** Hoopoe (top left), Little Owl (bottom left); Great White Pelican, Glossy Ibis, Stonechat, Pied Wheatear, Icterine Warbler

**BEST PERIOD** April–June; September–October

**ACCESS** Entrance in Belgorod town, at 46.2000°N, 30.3521°E. The entry charge to the fort itself is about €2, and is worth it just to climb the stairs to the top of the walls for excellent views of the liman.

# ⑭ Solone Lake

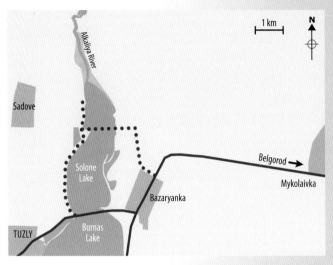

STATUS Part of Tuzly Lagoons National Nature Park, Ramsar Site, Important Bird Area 087

AREA ca. 820 ha

FEATURES This brackish water lake comprises the upper part of Burnas Lake, with a small freshwater inflow from the Alkaliya River. It has shallow water and silty margins with some low halophytic vegetation. The surrounding area is arable or steppe (with spotted suslik colonies). It is used as a resting place by gulls and a feeding area by migratory waders.

GOOD FOR Stone Curlew (bottom left), Curlew Sandpiper (top left); Pied Avocet, , Kentish Plover, Black-tailed Godwit, Slender-billed Gull

BEST PERIOD April–June; September–October

ACCESS The lake is cut by two causeways. The upper one is a track that joins Bazaryanka with Sadove and is not suitable for cars. The lower one is the road from Belgorod to Tuzly and Tatarbunary. There is a dirt track along the western side of the lake from 45.8797° N, 30.1166° E.

# ⑮ Khadzhyder Lake

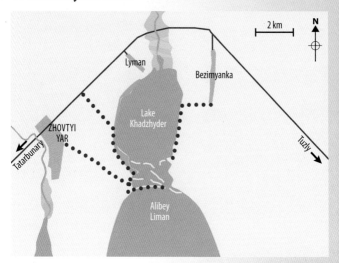

STATUS In Tuzly Lagoons National Nature Park, Ramsar Site, Important Bird Area 087
AREA ca. 2,000 ha

FEATURES Khadzhyder Lake forms an upper arm of the much larger Alibey Liman,
receiving a small inflow from the Khadzhyder River. It is mainly freshwater, with an
extensive reedbed at the top, and a complex of sandbars at the lower end which mark
it off from Alibey proper. These sand bars are used for breeding by various gull and
tern species, and waders use the silty margins. It is overlooked by a cliff that gives
good views over the lake, and some fringing shelter belts provide cover for passerines
and raptors.

GOOD FOR Pallas's Gull (top left), Black-tailed Godwit (bottom left); Shelduck, Pied
  Avocet, Collared Pratincole, Mediterranean Gull, Sandwich Tern

BEST PERIOD April–June; September–October

ACCESS The road from Tatarbunary to Tuzly crosses the reedbeds on a causeway at
  45.9167°N, 29.9866°E. A dirt track leading to the lower part of the site goes from
  Zhovtyi Yar starts at 45.8650°N, 29.9337°E, and can be followed
  back to the road at 45.8857°N, 29.9382°E.

# ⑯ Sasyk Liman, Tatarbunary

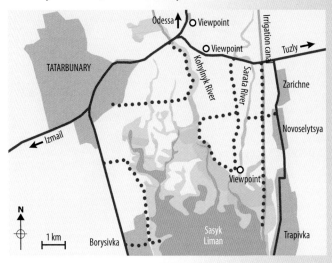

**STATUS** Ramsar Site, Important Bird Area 085

**AREA** ca. 2,180 ha

**FEATURES** This site lies at the junction of two broad valleys where the Rivers Kohylnyk and Sarata debouch, forming an extensive area of streams, pools, reedbeds and pastures. The Sarata River was canalised during the 1970s as part of an irrigation scheme and most of its course is now choked with reed. Farther south, the wetlands transition to the shallow open waters of Sasyk Liman proper.

**GOOD FOR** Crested Lark (top left), Great White Egret (bottom left); Great White Pelican, Purple Heron, Black Stork, Marsh Harrier, Hen Harrier, Red-footed Falcon, Quail, Black-winged Stilt

**BEST PERIOD** April–June; September–October

**ACCESS** The Odessa–Tatarbunary road (M15) passes the site; a viewpoint at 45.8532N, 29.6588E (next to the winery) overlooks the Kohylnyk valley and there are tracks leading to the marshes at 45.8481°N, 29.6513°E and 45.8215°N, 29.6208°E. The Sarata valley can be viewed from 45.8466°N, 29.6620°E, and a dirt track to the promontory is at 45.8433°N, 29.6794°E. The canal embankment can also be accessed at 45.8420°N, 29.6909°E.

## ⑰ Sasyk Liman, Primorske

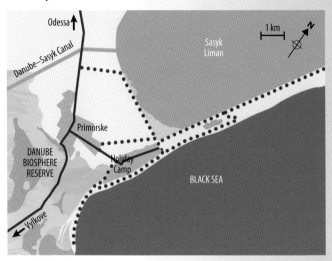

**STATUS** Ramsar Site, Important Bird Area 085

**AREA** ca. 2,015 ha

**FEATURES** This site comprises the coast and shore by Sasyk Liman, with dunes, brackish pools, reed patches and scrubby areas of sea buckthorn. Sasyk itself is isolated from the Black Sea by a concrete dam built in the 1980s. The site location concentrates migrants and the bay is good for seabirds, as well as harbour porpoise *Phocaena phocaena*.

**GOOD FOR** Red-breasted Goose (top left), Caspian Gull (bottom left); Great White Pelican, Whooper Swan, Smew, Peregrine Falcon, Pied Avocet, Kentish Plover, Black-tailed Godwit, Little Tern, Tawny Pipit, Red-throated Pipit

**BEST PERIOD** April–June; September–October

**ACCESS** The site is approached from the Spasky–Vylkove road through Primorske. There is a side road in the village at 45.5197°N, 29.6028°E leading to a holiday camp; outside of the summer season, driving through the camp is the quickest way to the coast, otherwise various tracks are available across the pasture areas. The dam can be driven, with care, along its whole length.

# 🔞 Vylkove: Danube Delta (by boat)

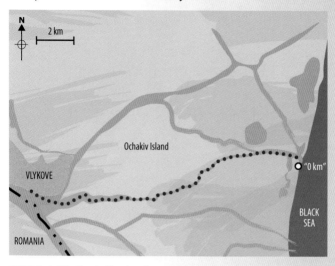

**STATUS** Part of the Danube Biosphere Reserve

**ROUTE DISTANCE** ca. 30 km

**FEATURES** The Kilia branch of the Danube forms the border between Ukraine and Romania. It now carries most of the flow and therefore most of the sediment, which means the Ukrainian side of the delta is still actively expanding (Vylkove was originally on the beach 250 years ago). The route follows the Ankudinove channel with its cultivated and wooded levees, and wide reedbeds, to the Black Sea coast. At the delta mouth is a lagoon which holds large numbers of waterbirds, and a monument to "0 km" (bottom left).

**GOOD FOR** Great White Pelican (top left); Dalmatian Pelican, Pygmy Cormorant, Little Bittern, Great Bittern, Spoonbill, Greylag Goose, Ferruginous Duck, White-tailed Eagle, Hobby, Kentish Plover, Black-tailed Godwit, Pallas's Gull, Gull-billed Tern, Caspian Tern, Kingfisher, Nightingale, Savi's Warbler, Icterine Warbler, Bearded Tit

**BEST PERIOD** April–October

**ACCESS** By boat from Vylkove. Tickets for tourist trips can be purchased from kiosks at the boat station (45.3977°N, 29.5815°E), while Salix Ecotours (see p.1) arranges special birding tours.

43

# ⑲ Vylkove: Zhebryanskyi Hryda

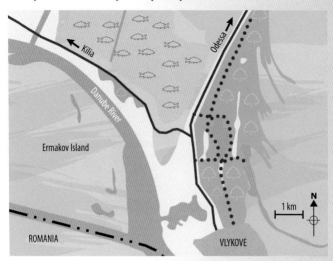

**STATUS** None

**AREA** ca. 500 ha

**FEATURES** This site is an arc of low dunes, some 5,000 years old, that marks the ancient Black Sea shoreline before Danube sediments filled in the bay. Originally, they were open and supported a unique flora. Unfortunately, the forest service planted them with pines in the 1980s, shading out much of the ground flora. However, the combination of dunes, slacks and trees attracts a good variety of birds.

**GOOD FOR** Roller (top left), Wood Lark (bottom left); Great White Pelican, Wryneck, Grey-headed Woodpecker, Tree Pipit, Hawfinch

**BEST PERIOD** April–June

**ACCESS** The Spasky–Vylkove road runs along the western edge of the site, and there is easy walking access (especially along the pylons). The only vehicle track starts at 45.4285°N, 29.5622°E; it leads to a lake (former sand quarry and nice for swimming) and on to the edge of some reedbeds.

# ㉒ Danube floodplain, Lisky

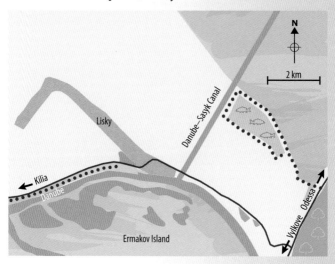

**STATUS** None

**AREA** ca. 3,890 ha

**FEATURES** The floodplain area between the Danube River and the extensive reedbeds to the north (Stentsivsko-Zhebriyanivski plavni) is one of the best general birding sites in Odessa. It includes fish ponds (filled or resting), riverine woodlands of mature poplars, reedbeds, pools, rice paddies and the river itself.

**GOOD FOR** Grey-headed Woodpecker (top left), Savi's Warbler (bottom left); Red-necked and Black-necked Grebe, Dalmatian Pelican, Pygmy Cormorant, Red-crested Pochard, Little Bittern, Black Woodpecker, Thrush Nightingale, Bluethroat, Marsh Warbler

**BEST PERIOD** April–September

**ACCESS** From the Vylkove–Spasky road, there is a raised concrete / dirt track at 45.4527°N, 29.5729°E that leads across reedbeds to the fish ponds next to the Sasyk-Danube canal. The lower part of the site is along the road to Lisky (and Kilia), with dirt track on to the flood embankment at 45.4578°N, 29.4840°E where there are several levee pools worth checking. The rice paddies, which attract waders, can be reached from 45.4589°N, 29.5007°E.

# ㉑ Aliaga steppe and Lake Kytai

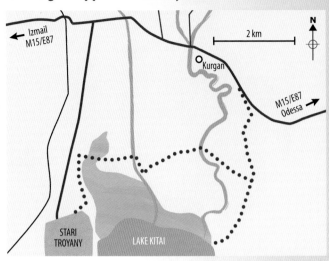

**STATUS** Important Bird Area 083

**AREA** ca. 2,020 ha

**FEATURES** The two steppic valleys at Aliaga, with the meandering small rivers of Aliaga (east) and Kyrhyzh-Kytai (west), provide an impressive vista of what the Bessarabian landscape looked like from ancient times to the 1960s when large-scale cultivation began. It is an important archaeological site for its many kurgans (burial mounds, top left) and a stretch of the Roman-era Traian's wall. The salt meadows in front of Lake Kytai provide feeding and roosting areas for thousands of wintering geese, while in summer the large rookery in the roadside windbreak supports dozens of nesting red-footed falcons.

**GOOD FOR** Red-footed Falcon (bottom left); Red-breasted Goose

**BEST PERIOD** January–February, May–August

**ACCESS** The Tatarbunary–Izmail road (M15) runs across the top of the site (caution: it is very busy). There is track to Kytai at 45.7175°N, 29.2188°E, and further exploration of the area can be done from the road to Stari Troyany at 45.7297°N, 29.1724°E.

# ㉒ Danube floodplain, Stara Nekrasovka to Kyslytsya

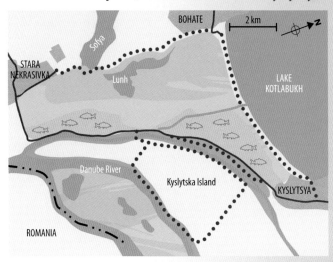

**STATUS** None

**AREA** ca. 2,900 ha

**FEATURES** This site holds the finest stand of easily accessible, regularly flooded, old-growth riverine forest in Odessa, with white and black poplar, willows and alder. There are adjacent reed beds (Lung reserve) and fish ponds which increase the bird interest of the area, while Kyslytska Island has more forest as well as arable land. When coming here, a short stop at the Struve Meridian Monument (a World Heritage Listed site, bottom left) is recommended.

**GOOD FOR** Sea Eagle (top left); Pygmy Cormorant, Spoonbill, White-tailed Eagle, Hobby, Kingfisher, Wryneck, Lesser Spotted Woodpecker, Black Woodpecker, Grey-headed Woodpecker, Thrush Nightingale, Nightingale, Marsh Warbler, Icterine Warbler

**BEST PERIOD** April–June

**ACCESS** The Izmail–Kilia road runs through the site. The main tracks are at 45.3728°N, 28.9792°E and 45.4038°N, 28.9504°E (to Lake Katlabugh shore, fish ponds, Bogate); 45.3504°N, 28.9282°E and 45.3965°N, 28.9450°E (to Safyani Lake, Lung and Katlabugh–Bogate); and 45.3652°N, 28.9754°E (ferry to Kyslytska island).

# ㉓ Isles of Izmail: Tataru Island (by boat)

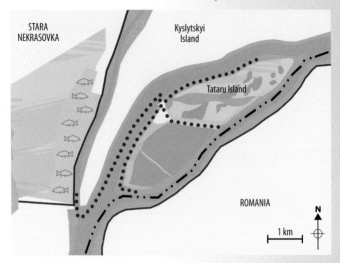

**STATUS** Part of Isles of Izmail Regional Landscape Park

**AREA** ca. 740 ha (Tataru)

**FEATURES** Tataru Island is an outstanding example of a river island – elongated canoe shape, a higher levee upstream than downstream (covered with pristine riverine forest of willow, poplar and alder, as well as oak on the highest parts), and a central depression with crystal clear open water containing a wealth of waterplants.

**GOOD FOR** Penduline Tit (top left), Red-necked Grebe (bottom left); Ferruginous Duck, Black Woodpecker, Bearded Tit, White-tailed Eagle, Hobby, White-winged Black Tern, Lesser Spotted Woodpecker, Thrush Nightingale

**BEST PERIOD** May–July

**ACCESS** The island is reached by a short boat trip (about 6 km) from the landing at Dunay, near Stara Nekrasovka: further information from Salix Ecotours (see p. 1). There are footpaths around and across the island, with a picnic area by the boat landing.

# ㉔ Izmail Fort

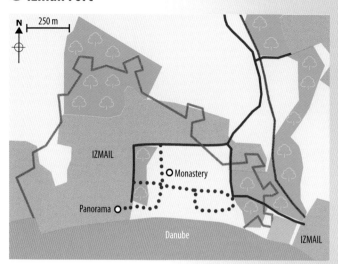

**STATUS** National Historical Monument

**AREA** ca. 270 ha

**FEATURES** This site lies in the grounds of a fort dating back to the 12th century that controlled the Danube navigation. It was the scene of many battles between Russian and Ottoman forces until Russian forces finally captured it in 1790 after a particularly brutal battle described by Lord Byron in *Don Juan*. The only building remaining is the mosque (bottom left), now converted to a panorama. The site gives a great view over the Danube, and the parkland holds a nice variety of woodland birds.

**GOOD FOR** Northern Wheatear (top left); Hobby, Wryneck, Syrian Woodpecker, Thrush Nightingale, Icterine Warbler

**BEST PERIOD** May–July

**ACCESS** The fort is situated southeast of the city centre, with its best entrance (with car park) at the mosque/panorama (45.3400°N, 28.8066°E).

# 25 Vynohradivka steppe reserve, Lake Bolgrad

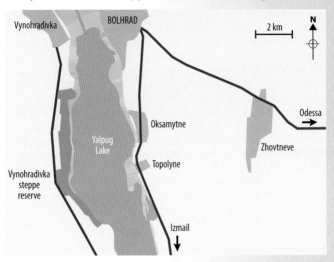

**STATUS** Botanical Nature Reserve

**AREA** ca. 420 ha

**FEATURES** The steppe reserve south of Vynohradivka stretches along a high cliff overlooking the western shore of Lake Bolgrad. Some of it has been planted with wretched stands of false acacia. There is a very rich, scented flora but the main interest for birding is the slip slopes and clefts filled with shrubby vegetation, where canopy species like golden oriole and ortolan bunting can be watched from above. It has a good population of Eastern Olivaceous Warblers nesting in the ravines.

**GOOD FOR** Ortolan Bunting (top left), Pied Wheatear (bottom left); Quail, Stock Dove, Tawny Pipit, Whinchat, Eastern Olivaceous Warbler, Golden Oriole, Raven

**BEST PERIOD** April–June

**ACCESS** The road from Reni via Kotlovyna to Bolhrad runs along the western side of the reserve. Vehicle access can be difficult as the field boundaries and ravine edges tend to change; the best track is at 45.6167°N, 28.5740°E. However, the site is not large and can be easily walked.

# ㉖ Lake Kurgurlui and Yalpug Causeway

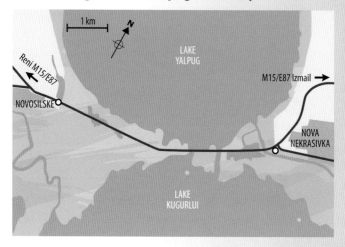

STATUS (Kugurlui only): Ramsar Site, Important Bird Area 081

CAUSEWAY LENGTH 5.5 km

FEATURES Lakes Yalpug (surface area 150 km²) and Kugurlui (90 km²) are the two largest natural freshwater lakes in Ukraine, and the core of the Lower Danube floodplain lakes system between Reni and Izmail. The bush-lined causeway is built on an old bank of the Danube. This is a top birding site as the causeway allows views over the wetlands, where waterbirds gather in enormous numbers, and terns and grebes nest close to it.

GOOD FOR Long-legged Buzzard (top left), Ferruginous Duck (bottom left); Dalmatian Pelican, Pygmy Cormorant, Red-crested Pochard, Honey Buzzard, Lesser Spotted Eagle, Booted Eagle, White-winged Tern, Black Tern, Penduline Tit

BEST PERIOD April–June; September–October

ACCESS The causeway links Izmail and Reni (M15) and is much used by trucks. It is possible to park and walk along the road verges, but parking is best at 45.3451°N, 28.6767°E or 45.3372°N, 28.6067°E. The latter is on a low elevation which lies under a migration route for soaring birds.

# ㉗ Lake Kartal and Danube River

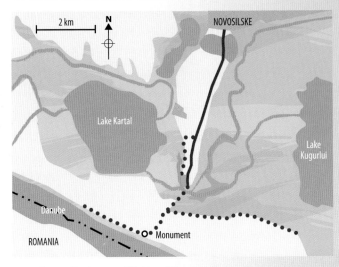

**STATUS** Ramsar Site, Important Bird Area 081

**AREA** ca. 560 ha

**FEATURES** This site is a low ridge across the Danube floodplain between Lakes Kugurlui and Kartal. It has muddy pools, open water, reedbeds and mature riverine forest. Polders either side of the track contain arable fields, weedy pasture and scrub. The ridge itself provides a migratory corridor for soaring birds.

**GOOD FOR** Black-winged Stilt (top left); Great White Pelican, Little Bittern, Glossy Ibis, Ferruginous Duck, White-tailed Eagle, Lesser Spotted Eagle, Grey-headed Woodpecker, Bluethroat, Spanish Sparrow

**BEST PERIOD** April–June; September–October

**ACCESS** The road to the site starts in Novosilske village at 45.3344°N, 28.5622°E; it leads to an angling club and then becomes a track to a riverside monument (bottom left) that marks where the Russian army under Nikolai I crossed the Danube in 1828 to fight the Turks. A fragment of the former Soviet electrified border fence also persists. It is not permitted to cross the fence into the forest. A track at 45.2865°N, 28.5452°E gives views over Lake Kartal – a gathering place for pelicans.

# ㉘ Lake Kagul, Orlivka

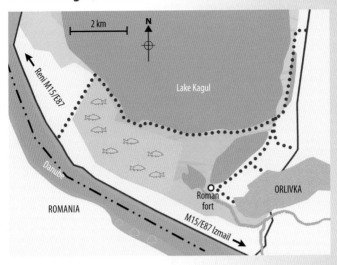

**STATUS** Important Bird Area 080

**AREA** ca. 1,490 ha (site)

**FEATURES** Lake Kagul is a large Danube floodplain waterbody of some 100 km² area just east of Reni that straddles the border with Moldova. This part of the lake has a cliffy eastern shore that gives wonderful views of the lake and floodplain against the backdrop of the Macin Mountains in Romania. It includes steppe, black poplar woodland, reedbeds and fish ponds. Just west of Orlivka is the site of Aliobrix Roman fort (bottom left).

**GOOD FOR** White Stork (top left); Whooper Swan, Red-breasted Goose, Ferruginous Duck, White-tailed Eagle, Pallid Harrier, Red-footed Falcon, Stonechat, Great Grey Shrike

**BEST PERIOD** April–June; September–October

**ACCESS** The road from Orlivka to Nahirne runs along the eastern side of the lake. A track from 45.3230°N, 28.4372°E to 45.3444°N, 28.4482°E covers the lakeside, and another at 45.3164°N, 28.4270°E leads to the Roman fort, which gives good views over the area.

# ㉙ Lake Kagul, Reni

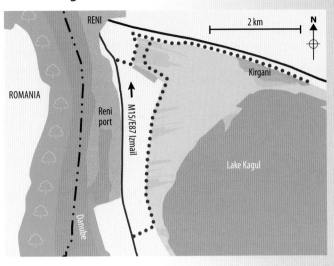

**STATUS** Important Bird Area 080

**AREA** ca. 500 ha

**FEATURES** Lake Kagul is a large Danube floodplain waterbody of some 100 km² area just east of Reni that straddles the border with Moldova. This part of the lake comprises reedbeds, with some open pools, and a clump of willows with a substantial colony of herons, Glossy Ibis and Pygmy Cormorants.

**GOOD FOR** Pygmy Cormorant (top left), Glossy Ibis (bottom left); Black-necked Grebe, Little Bittern, Great Bittern, Spoonbill, Greylag Goose, Ferruginous Duck, White-tailed Eagle, Lesser Spotted Eagle, River Warbler, Moustached Warbler, Marsh Warbler, Bearded Tit, Penduline Tit, Spanish Sparrow

**BEST PERIOD** April –June; September–October

**ACCESS** The site is accessible from a track off the Reni–Izmail road at 45.4339°N, 28.2944°E: this leads to a factory where reed is processed for thatch and biomass pellets. Turn left here and follow the track to the railway. Turn right at 45.4370°N 28.3034°E to follow the lake embankment (car access possible), or continue straight on to the small village of Kirgani (no car access after the railway siding).

# Checklist

The following checklist comprises the bird species most likely to be seen in Odessa province during the year – some are quite rare, but very rare or vagrant species are not included. Sites where certain species occur regularly are noted; absence of suggested sites indicates either a widespread bird, or one with unpredictable occurrence.

| ✔ | Name | Status | Suggested sites |
|---|------|--------|-----------------|
| | **Black-throated Diver** *Gavia arctica* | Scarce visitor | 5, 9 |
| | **Little Grebe** *Tachybaptus ruficollis* | Common resident | |
| | **Red-necked Grebe** *Podiceps grisegena* | Scarce resident | 12, 20, 23, 26 |
| | **Great Crested Grebe** *Podiceps cristatus* | Common resident | |
| | **Black-necked Grebe** *Podiceps nigricollis* | Common resident | 7, 9, 12, 17, 18, 20, 26, 29 |
| | **Levantine Shearwater** *Puffinus yelkouan* | Scarce passage | 9, 17 |
| | **Great White Pelican** *Pelecanus onocrotalus* | Scarce visitor | 5, 12, 13, 17, 18, 19, 26, 27 |
| | **Dalmatian Pelican** *Pelecanus crispus* | Rare visitor | 18, 20, 26, 27 |
| | **Great Cormorant** *Phalacrocorax carbo* | Common resident | |
| | **Pygmy Cormorant** *Phalacrocorax pygmeus* | Scarce resident | 11, 12, 18, 20, 22, 26, 27, 28, 29 |
| | **Grey Heron** *Ardea cinerea* | Scarce resident | |
| | **Purple Heron** *Ardea purpurea* | Common breeder | |
| | **Great Egret** *Ardea alba* | Common breeder | |
| | **Little Egret** *Egretta garzetta* | Scarce breeder | |
| | **Squacco Heron** *Ardeola ralloides* | Scarce breeder | |
| | **Night Heron** *Nycticorax nycticorax* | Common resident | |
| | **Little Bittern** *Ixobrychus minutus* | Scarce breeder | 7, 11, 12, 17, 18, 20, 26, 27, 29 |
| | **Great Bittern** *Botaurus stellaris* | Scarce breeder | 11, 12, 18, 27, 29 |
| | **Black Stork** *Ciconia nigra* | Scarce passage | 11, 26, 27, 29 |
| | **White Stork** *Ciconia ciconia* | Common breeder | |
| | **Glossy Ibis** *Plegadis falcinellus* | Scarce breeder | 11, 13, 26, 27, 28, 29 |
| | **Spoonbill** *Platalea leucorodia* | Rare breeder | 18, 22, 29 |
| | **Mute Swan** *Cygnus olor* | Scarce resident | |
| | **Whooper Swan** *Cygnus cygnus* | Scarce visitor | 17, 22, 28 |
| | **Greater White-fronted Goose** *Anser albifrons* | Common visitor | |
| | **Greylag Goose** *Anser anser* | Scarce resident | 18, 26, 29 |
| | **Red-breasted Goose** *Branta ruficollis* | Rare visitor | 11, 17, 21, 28 |
| | **Shelduck** *Tadorna tadorna* | Common resident | 4, 5, 7, 15, 17 |
| | **Wigeon** *Anas penelope* | Common visitor | |
| | **Gadwall** *Anas strepera* | Scarce visitor | |
| | **Teal** *Anas crecca* | Common visitor | |
| | **Mallard** *Anas platyrhynchos* | Common resident | |
| | **Pintail** *Anas acuta* | Rare visitor | |
| | **Garganey** *Anas querquedula* | Scarce breeder | |
| | **Shoveler** *Anas clypeata* | Scarce resident | |
| | **Red-crested Pochard** *Netta rufina* | Scarce resident | 12, 22, 26 |
| | **Pochard** *Aythya ferina* | Common resident | |
| | **Ferruginous Duck** *Aythya nyroca* | Scarce breeder | 12, 18, 23, 26, 27, 28, 29 |
| | **Tufted Duck** *Aythya fuligula* | Scarce visitor | |
| | **Eider** *Somateria mollissima* | Rare visitor | 9 |
| | **Goldeneye** *Bucephala clangula* | Scarce visitor | |
| | **Smew** *Mergellus albellus* | Scarce visitor | 17 |
| | **Red-breasted Merganser** *Mergus serrator* | Scarce visitor | 9, 17 |
| | **Osprey** *Pandion haliaetus* | Scarce passage | |
| | **Honey Buzzard** *Pernis apivorus* | Scarce breeder | 1, 26 |
| | **Black Kite** *Milvus migrans* | Scarce passage | 1 |
| | **White-tailed Eagle** *Haliaeetus albicilla* | Scarce resident | 5, 11, 18, 22, 26, 27, 28, 29 |
| | **Marsh Harrier** *Circus aeruginosus* | Common breeder | |
| | **Hen Harrier** *Circus cyaneus* | Scarce visitor | |
| | **Pallid Harrier** *Circus macrourus* | Rare passage | 27, 28 |
| | **Montagu´s Harrier** *Circus pygargus* | Scarce passage | 4, 5, 6 |
| | **Sparrowhawk** *Accipiter nisus* | Common visitor | |

| ✔ | Name | Status | Suggested sites |
|---|------|--------|-----------------|
| | **Goshawk** *Accipiter gentilis* | Rare resident | |
| | **Buzzard** *Buteo buteo* | Scarce breeder | 1, 3 |
| | **Long-legged Buzzard** *Buteo rufinus* | Scarce visitor | |
| | **Rough-legged Buzzard** *Buteo lagopus* | Rare visitor | |
| | **Lesser Spotted Eagle** *Aquila pomarina* | Scarce passage | 26, 27, 29 |
| | **Greater Spotted Eagle** *Aquila clanga* | Rare passage | |
| | **Booted Eagle** *Hieraaetus pennatus* | Scarce passage | 1, 26 |
| | **Kestrel** *Falco tinnunculus* | Common resident | |
| | **Red-footed Falcon** *Falco vespertinus* | Common breeder | 5, 16, 21, 28, 29 |
| | **Merlin** *Falco columbarius* | Scarce visitor | |
| | **Hobby** *Falco subbuteo* | Common breeder | 8, 9, 10, 12, 18, 22, 24 |
| | **Peregrine Falcon** *Falco peregrinus* | Scarce visitor | 17 |
| | **Grey Partridge** *Perdix perdix* | Common resident | |
| | **Quail** *Coturnix coturnix* | Common breeder | |
| | **Pheasant** *Phasianus colchicus* | Common resident | |
| | **Water Rail** *Rallus aquaticus* | Scarce resident | 18, 26, 29 |
| | **Spotted Crake** *Porzana porzana* | Rare breeder | 18 |
| | **Moorhen** *Gallinula chloropus* | Common resident | |
| | **Coot** *Fulica atra* | Common resident | |
| | **Oystercatcher** *Haematopus ostralegus* | Rare breeder | 17, 18 |
| | **Black-winged Stilt** *Himantopus himantopus* | Common breeder | |
| | **Pied Avocet** *Recurvirostra avosetta* | Scarce breeder | 5, 7, 14, 15, 17 |
| | **Stone Curlew** *Burhinus oedicnemus* | Scarce breeder | 14, 17, 18 |
| | **Collared Pratincole** *Glareola pratincola* | Rare breeder | 15 |
| | **Lapwing** *Vanellus vanellus* | Common breeder | |
| | **Golden Plover** *Pluvialis apricaria* | Scarce passage | 5 |
| | **Grey Plover** *Pluvialis squatarola* | Scarce passage | 5, 17, 18 |
| | **Ringed Plover** *Charadrius hiaticula* | Scarce passage | 5, 17 |
| | **Little Ringed Plover** *Charadrius dubius* | Scarce breeder | 5, 17 |
| | **Kentish Plover** *Charadrius alexandrinus* | Scarce breeder | 5, 14, 17, 18 |
| | **Woodcock** *Scolopax rusticola* | Scarce passage | 19 |
| | **Snipe** *Gallinago gallinago* | Common passage | |
| | **Black-tailed Godwit** *Limosa limosa* | Scarce passage | 5, 14, 17, 18 |
| | **Curlew** *Numenius arquata* | Scarce visitor | 5, 17, 18 |
| | **Terek Sandpiper** *Xenus cinereus* | Rare passage | 5 |
| | **Common Sandpiper** *Actitis hypoleucos* | Scarce breeder | |
| | **Green Sandpiper** *Tringa ochropus* | Scarce passage | |
| | **Spotted Redshank** *Tringa erythropus* | Common passage | |
| | **Greenshank** *Tringa nebularia* | Common passage | |
| | **Marsh Sandpiper** *Tringa stagnatilis* | Scarce passage | 5 |
| | **Wood Sandpiper** *Tringa glareola* | Common passage | |
| | **Redshank** *Tringa totanus* | Scarce breeder | |
| | **Ruddy Turnstone** *Arenaria interpres* | Scarce passage | 5, 17 |
| | **Sanderling** *Calidris alba* | Rare passage | |
| | **Little Stint** *Calidris minuta* | Common passage | |
| | **Temminck's Stint** *Calidris temminckii* | Scarce passage | 5 |
| | **Curlew Sandpiper** *Calidris ferruginea* | Common passage | |
| | **Dunlin** *Calidris alpina* | Common passage | |
| | **Broad-billed Sandpiper** *Limicola falcinellus* | Scarce passage | 5 |
| | **Ruff** *Philomachus pugnax* | Common passage | |
| | **Red-necked Phalarope** *Phalaropus lobatus* | Scarce passage | 5 |
| | **Arctic Skua** *Stercorarius parasiticus* | Rare visitor | 9 |
| | **Slender-billed Gull** *Chroicocephalus genei* | Scarce breeder | 5, 14 |
| | **Black-headed Gull** *Chroicocephalus ridibundus* | Common resident | |
| | **Little Gull** *Hydrocoloeus minutus* | Common passage | 5 |
| | **Western Yellow-legged Gull** *Larus michahellis* | Scarce visitor | |
| | **Lesser Black-backed Gull** *Larus fuscus* | Scarce visitor | 9, 17 |
| | **Caspian Gull** *Larus cachinnans* | Common resident | |
| | **Common Gull** *Larus canus* | Common visitor | |
| | **Mediterranean Gull** *Ichthyaetus melanocephalus* | Common passage | 5, 15 |

| ✔ | Name | Status | Suggested sites |
|---|------|--------|-----------------|
| ☐ | **Pallas's Gull** *Ichthyaetus ichthyaetus* | Rare breeder | 15, 18 |
| ☐ | **Little Tern** *Sternula albifrons* | Scarce breeder | 5, 17, 18 |
| ☐ | **Gull-billed Tern** *Gelochelidon nilotica* | Rare breeder | 18 |
| ☐ | **Caspian Tern** *Hydroprogne caspia* | Rare breeder | 18 |
| ☐ | **Whiskered Tern** *Chlidonias hybrida* | Common breeder | |
| ☐ | **White-winged Tern** *Chlidonias leucopterus* | Rare breeder | 26, 27, 29 |
| ☐ | **Black Tern** *Chlidonias niger* | Scarce breeder | 26 |
| ☐ | **Common Tern** *Sterna hirundo* | Common breeder | |
| ☐ | **Sandwich Tern** *Thalasseus sandvicensis* | Scarce breeder | 9, 17, 18 |
| ☐ | **Stock Dove** *Columba oenas* | Scarce passage | 25 |
| ☐ | **Wood Pigeon** *Columba palumbus* | Common resident | |
| ☐ | **Turtle Dove** *Streptopelia turtur* | Common breeder | |
| ☐ | **Collared Dove** *Streptopelia decaocto* | Common breeder | |
| ☐ | **Cuckoo** *Cuculus canorus* | Common breeder | |
| ☐ | **Little Owl** *Athene noctua* | Common resident | |
| ☐ | **Long-eared Owl** *Asio otus* | Scarce resident | |
| ☐ | **Nightjar** *Caprimulgus europaeus* | Scarce breeder | |
| ☐ | **Swift** *Apus apus* | Common breeder | |
| ☐ | **Kingfisher** *Alcedo atthis* | Scarce resident | 18, 22, 29 |
| ☐ | **Bee-eater** *Merops apiaster* | Common breeder | |
| ☐ | **Roller** *Coracias garrulus* | Common breeder | |
| ☐ | **Hoopoe** *Upupa epops* | Common breeder | |
| ☐ | **Wryneck** *Jynx torquilla* | Scarce breeder | 1, 3, 8, 9, 12, 18, 19, 22, 24 |
| ☐ | **Lesser Spotted Woodpecker** *Dendrocopos minor* | Rare breeder | 22 |
| ☐ | **Middle Spotted Woodpecker** *Dendrocopos medius* | Rare breeder | 1, 27 |
| ☐ | **Great Spotted Woodpecker** *Dendrocopos major* | Scarce resident | 1, 3 |
| ☐ | **Syrian Woodpecker** *Dendrocopos syriacus* | Common resident | |
| ☐ | **Black Woodpecker** *Dryocopus martius* | Rare resident | 18, 20, 22, 23 |
| ☐ | **Grey-headed Woodpecker** *Picus canus* | Scarce resident | 18, 20, 22, 27, 29 |
| ☐ | **Calandra Lark** *Melanocorypha calandra* | Rare resident | 4 |
| ☐ | **Short-toed Lark** *Calandrella brachydactyla* | Scarce resident | |
| ☐ | **Crested Lark** *Galerida cristata* | Common resident | |
| ☐ | **Skylark** *Alauda arvensis* | Common resident | |
| ☐ | **Wood Lark** *Lullula arborea* | Rare breeder | 4, 19 |
| ☐ | **Sand Martin** *Riparia riparia* | Scarce breeder | |
| ☐ | **Swallow** *Hirundo rustica* | Common breeder | |
| ☐ | **House Martin** *Delichon urbicum* | Common breeder | |
| ☐ | **Tawny Pipit** *Anthus campestris* | Common breeder | 5, 17, 25 |
| ☐ | **Meadow Pipit** *Anthus pratensis* | Scarce visitor | |
| ☐ | **Red-throated Pipit** *Anthus cervinus* | Scarce passage | 4, 5, 17 |
| ☐ | **Tree Pipit** *Anthus trivialis* | Scarce breeder | 4, 19 |
| ☐ | **White Wagtail** *Motacilla alba* | Common breeder | |
| ☐ | **Black-headed Wagtail** *Motacilla flava feldegg* | Common breeder | |
| ☐ | **Citrine Wagtail** *Motacilla citreola* | Rare passage | 5 |
| ☐ | **Grey Wagtail** *Motacilla cinerea* | Rare passage | |
| ☐ | **Wren** *Troglodytes troglodytes* | Scarce visitor | 9, 27, 29 |
| ☐ | **Mistle Thrush** *Turdus viscivorus* | Scarce visitor | 1 |
| ☐ | **Song Thrush** *Turdus philomelos* | Common passage | |
| ☐ | **Redwing** *Turdus iliacus* | Rare visitor | |
| ☐ | **Fieldfare** *Turdus pilaris* | Scarce visitor | |
| ☐ | **Blackbird** *Turdus merula* | Common resident | |
| ☐ | **Thrush Nightingale** *Luscinia luscinia* | Common breeder | 1, 9, 11, 12, 18, 19, 20, 22, 24 |
| ☐ | **Nightingale** *Luscinia megarhynchos* | Scarce breeder | 18, 22 |
| ☐ | **Bluethroat** *Luscinia Scarce visitorecica* | Rare breeder | 11, 20, 27 |
| ☐ | **Robin** *Erithacus rubecula* | Scarce visitor | 1, 3, 9 |
| ☐ | **Black Redstart** *Phoenicurus ochruros* | Common breeder | |
| ☐ | **Redstart** *Phoenicurus phoenicurus* | Common breeder | |
| ☐ | **Whinchat** *Saxicola rubetra* | Scarce breeder | 9, 25 |
| ☐ | **Stonechat** *Saxicola rubicola* | Scarce breeder | 13, 28 |
| ☐ | **Wheatear** *Oenanthe oenanthe* | Common breeder | |

| ✔ | Name | Status | Suggested sites |
|---|------|--------|-----------------|
| | **Pied Wheatear** *Oenanthe pleschanka* | Scarce breeder | 13, 25 |
| | **Isabelline Wheatear** *Oenanthe isabellina* | Rare breeder | 4, 6 |
| | **Spotted Flycatcher** *Muscicapa striata* | Common breeder | |
| | **Pied Flycatcher** *Ficedula hypoleuca* | Scarce passage | 3, 5, 8, 9, 10, 18 |
| | **Collared Flycatcher** *Ficedula albicollis* | Rare breeder | 9, 10, 18, 29 |
| | **Red-breasted Flycatcher** *Ficedula parva* | Scarce passage | 5, 9, 18, 19 |
| | **Goldcrest** *Regulus regulus* | Scarce visitor | 9, |
| | **Savi's Warbler** *Locustella luscinioides* | Common breeder | 11, 18, 20 |
| | **River Warbler** *Locustella fluviatilis* | Rare breeder | 29 |
| | **Moustached Warbler** *Acrocephalus melanopogon* | Rare breeder | 29 |
| | **Sedge Warbler** *Acrocephalus schoenobaenus* | Common breeder | |
| | **Paddyfield Warbler** *Acrocephalus agricola* | Scarce breeder | 5, 11, 18, 20 |
| | **Reed Warbler** *Acrocephalus scirpaceus* | Common breeder | |
| | **Marsh Warbler** *Acrocephalus palustris* | Common breeder | 20, 22, 29 |
| | **Great Reed Warbler** *Acrocephalus arundinaceus* | Common breeder | |
| | **Eastern Olivaceous Warbler** *Iduna pallida* | Rare breeder | 25, 29 |
| | **Icterine Warbler** *Hippolais icterina* | Scarce breeder | 13, 18, 22, 24, 27, 29 |
| | **Willow Warbler** *Phylloscopus trochilus* | Scarce passage | |
| | **Chiffchaff** *Phylloscopus collybita* | Common passage | |
| | **Wood Warbler** *Phylloscopus sibilatrix* | Scarce passage | 1, 3 |
| | **Blackcap** *Sylvia atricapilla* | Common breeder | |
| | **Garden Warbler** *Sylvia borin* | Scarce passage | |
| | **Barred Warbler** *Sylvia nisoria* | Scarce breeder | 29 |
| | **Lesser Whitethroat** *Sylvia curruca* | Common breeder | |
| | **Whitethroat** *Sylvia communis* | Common breeder | |
| | **Bearded Tit** *Panurus biarmicus* | Scarce resident | 7, 11, 12, 17, 18, 23, 27, 29 |
| | **Marsh Tit** *Poecile palustris* | Rare visitor | |
| | **Coal Tit** *Parus ater* | Scarce visitor | |
| | **Great Tit** *Parus major* | Common resident | |
| | **Blue Tit** *Cyanistes caeruleus* | Common resident | |
| | **Penduline Tit** *Remiz pendulinus* | Scarce breeder | 11, 12, 23, 26, 29 |
| | **Long-tailed Tit** *Aegithalos caudatus* | Scarce visitor | |
| | **Treecreeper** *Certhia familiaris* | Scarce visitor | 1, 3, 9 |
| | **Golden Oriole** *Oriolus oriolus* | Common breeder | |
| | **Red-backed Shrike** *Lanius collurio* | Common breeder | |
| | **Lesser Grey Shrike** *Lanius minor* | Common breeder | |
| | **Great Grey Shrike** *Lanius excubitor* | Rare visitor | 11, 28 |
| | **Jay** *Garrulus glandarius* | Scarce resident | 1, 3, 9 |
| | **Magpie** *Pica pica* | Common resident | |
| | **Nutcracker** *Nucifraga caryocatactes* | Rare visitor | |
| | **Jackdaw** *Corvus monedula* | Common resident | |
| | **Rook** *Corvus frugilegus* | Common resident | |
| | **Hooded Crow** *Corvus cornix* | Common resident | |
| | **Raven** *Corvus corax* | Scarce resident | 1, 3, 6, 25 |
| | **Starling** *Sturnus vulgaris* | Common resident | |
| | **Rose-coloured Starling** *Pastor roseus* | Rare passage | |
| | **House Sparrow** *Passer domesticus* | Common resident | |
| | **Spanish Sparrow** *Passer hispaniolensis* | Scarce breeder | 27, 29 |
| | **Tree Sparrow** *Passer montanus* | Common resident | |
| | **Chaffinch** *Fringilla coelebs* | Common resident | |
| | **Brambling** *Fringilla montifringilla* | Scarce visitor | |
| | **Greenfinch** *Chloris chloris* | Common resident | |
| | **Goldfinch** *Carduelis carduelis* | Common resident | |
| | **Linnet** *Carduelis cannabina* | Scarce resident | |
| | **Bullfinch** *Pyrrhula pyrrhula* | Scarce visitor | 19 |
| | **Hawfinch** *Coccothraustes coccothraustes* | Scarce resident | |
| | **Corn Bunting** *Emberiza calandra* | Common resident | |
| | **Yellowhammer** *Emberiza citrinella* | Scarce visitor | 1 |
| | **Ortolan Bunting** *Emberiza hortulana* | Common breeder | |
| | **Reed Bunting** *Emberiza schoeniclus* | Common resident | 5, 11, 12, 18, 26, 27, 29 |